Careers in FASHION & CLOTHING DESIGN

your questions and answers

SECOND EDITION

TROTMAN

This second edition published in 2000 in Great Britain
by Trotman and Company Limited,
2 The Green, Richmond, Surrey TW9 1PL
First edition published in 1995

British Library Cataloguing in Publication Data
A catalogue record for this book is available from the British Library
ISBN 0 85660 562 X

Printed and bound in Great Britain
by Creative Print & Design (Wales) Ltd

Contents

Acknowledgements

The publishers would like to thank Tony Charlton of UCAS for his help updating this new edition. CAPITB Trust also provided useful statistics and information on work and training in the clothing industry.

What about working in fashion & clothing design?

You can probably name half a dozen famous designers, not to mention the supermodels who glide up and down the catwalks in their creations. You've seen the glossy fashion magazines that every season show us what is happening on the catwalks of the world. You may even have been to the 'Clothes Show Live'. The shops up and down the country where we buy our clothes inspire the same allegiances in us as our favourite bands. What you probably don't know is that clothes making is the one of the largest manufacturing industries in Britain, employing around 165,000 people, and with textiles it produces around £16 billion worth of goods a year.

There are around 2000 fashion designers in Britain – they are not merely creating our clothes, but the whole way that we see ourselves and our concepts of what is attractive, appropriate and modern. Fashion and clothing designers are very aware of current trends and moods in society, the music business and the media. This has created a range of specialist jobs, such as colour predictors and fashion forecasters who shape the fashions that we will be wearing in a year or two years' time. While fashion designers are to be found in every type of clothing company, they are only a small proportion of the many workers in this industry, some of whose jobs are also outlined in this book.

The fashion industry is commonly divided into three main sectors, each with different methods of working, different clientele and, significantly, very different price ranges for the clothes they sell. While there is inevitably some blurring between these categories, it will help you to think in terms of these three sectors to be aware of the career possibilities in fashion design.

1. Haute couture fashion

The area of fashion which is the most prestigious and exclusive is known as 'haute couture' – a French term that translates into English as 'high fashion'. Haute couturiers are concerned with creating exclusive, individual clothes, made to measure, for a small number of individual clients. The price of these garments can run into thousands of pounds, which is a reflection of the quality of the materials used, the personal service, the exclusivity of the designs, the production methods (the most traditional in the business) and the fact that these clothes are on the cutting edge of fashion.

Haute couture accounts for only a small number of the total designs created by fashion designers, but these can be the most influential and trend-setting, the most outlandish and unusual creations in the business.

2. Ready to wear fashion

The middle range of the clothing industry is occupied by a sector variously known as 'off the peg', 'ready to wear' or, confusingly enough, 'designer fashion'. These clothes and accessories are created by prestigious, often well-known designers, and are recognised by their label (eg Tony Choo shoes, a Karen Millen dress, Prada sunglasses). They are made in quantities in factories and then sold through shops, exclusive boutiques and 'designer' chains.

High-quality fabrics and very fashionable, wearable designs characterise these clothes. The stores and the clothes themselves have a luxury air. This is a growing sector of the market, as more and more people are buying a few special occasion clothes from these stores to supplement a wardrobe made up of less expensive items.

3. Mass market fashion

'High street' or 'mass market' fashion is the name applied to the clothes that most of us wear every day and, not surprisingly, this sector is by far the largest in the industry. Diversity, huge production volumes and safe bets characterise high street fashion. The designs of the clothes are mostly adapted from the current fashionable styles rather than setting trends themselves. They are sold through the stores that are found in high streets up and down the country, such as Top Shop, Burtons and The Gap.

Bear in mind that these distinctions are general; not all fashion companies can be slotted neatly into one of the above categories. The differences between the sectors are becoming less pronounced as companies try to move into gaps in the market; for example, many designers now offer a 'diffusion' line which is produced for the mass market. Also, bespoke tailoring, which is the production of high-quality, hand-crafted clothes for individual customers, is a separate area that doesn't really lead fashion, but is at the top end of the market and requires very highly skilled professionals.

What sorts of jobs are there?

Fashion design

As outlined in the previous section, there are three areas of fashion design – mass market, ready to wear and haute couture. In fact, most people in the clothing industry work in manufacturing, as machinists and cutters, as well as pattern makers, pressers, distribution staff, and so on. In this book we look at the process as starting from garment design. (For more information on careers in manufacturing, contact CAPITB Trust, which is the national training organisation for the clothing industry.) The designers and design teams in each area work as follows.

Mass market design

The design department in a mass market fashion company is an extremely busy place. It is standard practice for each person to work with one category of clothing, such as accessories or men's shirts. Coordinating the work of this team will be a senior designer, who makes sure that the company's collection is cohesive and who works out the budgets for each category of the collection with the financial department. As these companies generally produce vast quantities of each garment, it is very important that each piece works in terms of look, coordination with other parts of the collection and budget.

When the next year's collection is in preparation, the senior designer meets with each individual designer in the team and they discuss the outline for the collection. The designers then begin work on their specific garment type. A great deal of planning is required for each garment, since it must be designed to fit within fairly narrow specifications.

Once the preparatory sketches have been made, fabrics must be found to make up the designs. At this point the designer performs a juggling act – because of the pressures of the budget, he or she must balance out the cost of the material with the complexity of the design. If the item will require a lot of sewing time, then the fabric must be cheaper, whereas if the design is simple with fewer seams, better quality fabric and fastenings can be afforded. Sometimes designers find this trade-off unsatisfactory and too limiting to their creativity. Once these initial plans are worked out, the designer again meets with the senior designer to discuss them, when any necessary modifications will be made.

Once the design is finalised, it goes to the sample room where a pattern is cut and a sample made up. If the sample meets the approval of the buyers, the garment's details are sent to the factory where the production process takes place.

It is worth briefly following the progress of the garment from this point. The pattern which was used for the sample is graded, which means that it is scaled up and down into the standard UK sizes (10–18+). The layout of the pattern pieces is determined in such a way that the maximum use is made of the fabric, with as little wastage as possible. The cutting and making up of the garment then begins. Once made up and finished the clothes are packed up and distributed via the warehouse to the stores, ultimately finding their way into our wardrobes.

Ready to wear design

The entire collection of a 'designer' store or chain is often conceived solely by one designer. This is a heavy workload for one person, but it also means that the work has a wide variety. Planning is still important, although not as rigid as for the mass market designer. Each season's collection is conceived a year in advance with the same detailed research and development going into the ideas.

The main person that the designer will consult is the sales manager who examines how successful the previous season's designs have been. The stages of preparation are generally very similar to those described above, with the design moving from the drawing board, to the 'story board' (a display of designs put together with fabric samples and a text description), to the sample room, to production.

In this sector the production itself may be done by small in-house factories or contracted out to clothing manufacturing companies. The designer's role in this sector of the market is far more fluid and influential. Whereas the mass market designer is following trends that are already in evidence, adapting the top designers' general silhouettes and styles, this type of designer is much more original, setting as well as following trends and catering for a narrower range of customers.

Haute couture design

The collections of these designers, which are shown at fashion shows, have a dual purpose. Clients can can choose the garments they like from the collection, to be tailor-made to their own measurements. These shows also catch the media's attention, thereby attracting more clients and boosting the designer's or the design house's image.

Haute couture is not just for the catwalks. There are many more small and little known companies who attract clients mainly by word of mouth. Designers who have set up their own small company will not only have to design clothes, but also manage the company, pay staff (such as a machinist, a pattern cutter and a junior) as well as paying their own salary, do the accounts, organise insurance and so on.

When clients make their first visit, the designer conducts a consultation with them. This meeting is to establish what the client wants, which is more difficult than it may sound. It is the job of the designer to steer the client tactfully towards specifics, as well as to clothes that are likely to suit them and look good. After all, couturiers' clothes are advertisements for their work, so it is important not to dress clients in outfits that don't suit them. After the initial consultation, the designer sets to work sketching and sourcing fabrics.

At the second meeting the client is presented with a selection of ideas to choose from and possible adaptations are considered. Money will also be discussed specifically at this point, the designer providing a price for the finished garment, which covers all the work and materials used.

The client is then measured, and from these measurements the designer makes up a dummy to the right shape. This is called a 'toile' and is created by wrapping a dressmaker's dummy with fine cloth until it is the same shape as the client. A pattern is cut for the garment by draping the toile with fabric, checking the hang of the fabric and adjusting the design.

The garment, lightly sewn, is then fitted on the client at intervals during the making-up processes and the finer points of the fit are modified. The sewing is done both by a machinist and also by hand – a strong emphasis on traditional methods and hand sewing characterises this sector of design. Finally, the customer is presented with an original piece of clothing which fits perfectly.

Tailoring

Tailoring companies can vary in size from small firms of a few people who are skilled in all aspects of the work, to larger companies where people might specialise in different areas, such as pattern cutting and stitching. As with haute couture, a customer has an initial consultation to have measurements taken and discuss style of garment and choice of fabric. The tailor quotes for the work and fabric, and if the price is agreed a pattern is made up to the customer's measurements, and the material is cut. This is work that requires skill and precision – tailored garments must fit very well and should be comfortable, but because they are made to the shape of each individual, there is the possibility of alterations at a later stage. After the first fitting, the garment is sewn using machines and hand stitching, pressed, and fitted again.

Savile Row, in London, is considered the world centre of excellence in bespoke tailoring, but the UK as a whole has a very good international reputation for tailoring – in both menswear and ladieswear – and in many large cities you will find high-quality tailors.

Fashion illustration

The drawings produced by fashion designers follow certain style conventions. These enable the designer to give an impression of the 'mood' of the garment while indicating technical detail. Fundamental to the design process, fashion illustrations also have applications outside the studio. While mainly photographs are used to show fashion designs in the media, illustration can also be used in magazines, books or newspapers. Illustration is also a vital part of fashion prediction (see below), as photography can't be used for garments that don't yet exist!

Illustrators are mainly freelance workers commissioned to undertake work for stores, advertising agencies and a variety of other employers. Pattern making is an area in which it is possible for the fashion illustrator to find a permanent position. Here they will work on covers and catalogue illustrations for paper patterns, as well as illustrating the instructions inside.

Fashion prediction and forecasting

Fashion prediction agencies employ freelance designers and illustrators. These agencies keep manufacturers, designers, merchandisers and retailers informed of the latest trends, not only in clothes but a range of items.

Although everyone in the fashion industry has to have a feeling for the way trends are developing, there are not many people working in this highly specialised field. Those who do specialise in this area spend a lot of time travelling and picking up on fashion and wider cultural, musical and political movements and moods around the world in order to put together portfolios of their predictions to sell to clients.

Fashion stylist

An understanding of how fabrics, textures and colours work together is a core skill of the fashion designer. If you also have a strong fashion sense, a tendency towards the dramatic and an ability to deal with highly strung people, you could become a successful fashion stylist. Stylists coordinate shoots for photographic or film projects. They put together the clothes and props to achieve a visual effect. Work is usually freelance and comes from magazines, television and video production companies, the music industry and PR companies.

Theatrical costume design

The designer who doesn't want to be confined to creating strictly wearable clothes may be interested in pursuing a career as a theatrical costume designer. A strong sense of dramatic effect, a very vivid imagination and a comprehensive understanding of the history of costume are important attributes for this work. Another key skill is the ability to design across the range of garments and accessories, as the designer will single-handedly provide all these items, whether by creating them or by sourcing them to hire or buy.

Costumes have a very powerful effect on the visual impact of a production. Consulting regularly with the producer of a play, film or video, the designer will decide how to create costumes for the players that will help bring out their characterisation and the overall interpretation of the production.

A costume designer in a small theatre (or other production) company will also be responsible for storing existing costumes, transporting them around, repairing and cleaning them. In a larger company these duties are shared among the wardrobe department and with a design assistant.

Fashion promotion

It's no good if a designer is creating the most wonderful clothes in the world, yet nobody knows about them. Fashion is above all else a business, and for a designer to be successful, he or she must be the name on everybody's lips. The promoter or PR (public relations) person has the responsibility for putting it there. Using contacts, charm and persuasion they attract publicity for their clients through media coverage and specially organised events. It is a glamorous career for the person with a combination of nerve, style, tact, relentless energy and a design or public relations background.

Fashion buying

The responsibility of the fashion buyer is to purchase the clothes for a store's collection. As volumes and budgets are very important, buyers must have a very keen commercial sense and strong awareness of the customer profile. Clothes may come from producers all over the world, so the job involves quite a lot of travel.

A buyer must also have the predictor's knack for forecasting what will sell in a year's time, as buying has to be done well in advance. Good planning and negotiating skills are essential as there are strict budgets to keep to while at the same time the best possible products must be found for the store and customer. A very keen fashion eye, coupled with retail experience, will equip you for this kind of job. The role is closely related to that of the fashion merchandiser.

Fashion merchandising

Also working with sales data and budgets is the fashion merchandiser. He or she looks at the clothes that the fashion buyer has bought or that the in-house design team has produced, and is responsible for deciding how, when and where they should be distributed around a chain or group of stores. The job is mostly desk based and any travelling is likely to be around the stores themselves, in order to assess the customer profile of each region. The store Jigsaw, for example, has 17 branches in London alone, each of which stocks part of the stores' range to suit the customer profile of that small area. A strong retail background is the most common passport into this work.

Colourist

If you have a particularly good eye for and interest in colour, you could consider a career as a colourist. The work has a strong predictive element, looking at trends in colour ranges. Colourists often come from a fine art background, where they have gained an excellent understanding of the finer points of tone and shade. These jobs are mainly within the large design studios, textile companies and dye manufacturers.

Fashion journalism

As well as fashion magazines, both tabloid and broadsheet papers run fashion articles. To become a fashion writer you need a trained background in either fashion design or journalism, or if you know early on that this is your chosen career, you could opt for one of the fashion journalism courses on offer. The danger with this last option is that you could find you are specialising too early on, which can be problematic if you discover that it is not really for you.

Journalism can be studied as a one-year postgraduate course once you have completed a design degree, although you are likely to have to fund this yourself; or training may sometimes be available within a newspaper group. It is a very exciting job, combining the glamour of visiting shows around the world with the practical skill of writing. Good interpersonal skills are important too, as you may need to work with a stylist to put together a fashion shoot to illustrate the article (this is more common in magazines). London is the base for most of these jobs, and you should be prepared for very stiff competition for a very limited number of jobs or commissions.

Fashion photography

The fashion photography of David Bailey shot him to fame and fortune in the 1960s. Since then, many photographers have looked to emulate his success through careers in the media industries (fashion, advertising, publishing, music, promos, etc). Although there are a small number of courses that specialise in fashion photography, many entrants to the profession have come from initially studying general photography, fashion promotion or, with far less frequency, visual communication or fine art. There is no doubt that London-based fashion photography students have the edge in terms of fashion exhibitions, events and possibilities for work placements.

Other jobs

Those described are just some of the many jobs in the fashion world that might use your art and design skills and training. There are a number of other career possibilities in clothing manufacture, fashion and textiles, that require different aptitudes and qualifications. If you are attracted to the world of fashion, but want to look at the wider possibilities, all these roles are needed in this industry:

- Marketing
- Retail (sales)
- Production management
- Stock ordering
- Textile design
- Sample room jobs – pattern cutter, machinist, sample room junior.

See page 40 for further reference sources, and contact CAPITB Trust, the national training organisation, for more information about jobs in the fashion industry.

What qualifications will I need?

Not all of the jobs described in the previous section require additional further education training in fashion design, but if your aspirations lie in becoming a designer, it is absolutely essential to gain a design qualification. The reason for this is that the competition for design jobs is so fierce that if you are not a graduate you simply do not stand a chance. If you are not prepared to undertake and survive the rigours of this tough training, a career as a designer is certainly not for you, and you should look more closely at some of the other jobs mentioned, or at another area of work altogether.

There are two main routes to higher education qualifications in fashion design. The first is the Edexcel/BTEC (or Scottish National awards) path, leading to a Higher National Diploma, the second is the art school/university route leading to a degree in fashion design, although it is possible to cross over between the two. The full course options at 18 are outlined on page 37; here we deal with the situation up to age 17/18.

The most common basis for these courses is GCSE qualifications. If you are considering a career in fashion you should obtain as many good passes (C+, 5 if possible) at this level as you can. Relevant subjects include art, home economics, textiles and design and technology, as well as English and maths which are considered the most basic qualifications for continuing education. Alternatively you can study for vocational qualifications in Art and Design, which are more job specific. The First Certificate and the Intermediate GNVQ are both roughly equivalent to four grade A–C GCSEs. A bonus of this route is that you are actively working towards what you want to do from an earlier point in your education;

the downside is that if you change your mind you have already limited your options.

At 16 you face a number of choices. The A-level route saves you from narrowing down your options too soon, by not reducing your range of subjects too early. Starting from Autumn 2000, most students will take four AS-levels in their first year of sixth-form study. To get on to an art foundation course you usually only require one A-level in art, but in reality course tutors will often be looking for two or even three passes. A strong portfolio is more important, however, and your art A-level should help you develop one of these.

Another option at 16 is to take a National Diploma in Art and Design or an Advanced GNVQ (or Scottish equivalents). These have the advantage of being vocationally oriented and much more practical than A-level qualifications.

For specialist work in manufacture, for example cutting, sewing, pressing and distribution, National (or Scottish) Vocational Qualifications (N/SVQs) give you work-based training. The National Traineeship prepares you for Level 2 N/SVQ. From this you can move on to Level 3 N/SVQ training through the Modern Apprenticeship. These schemes have been developed by CAPITB Trust, who you should contact for more information on training opportunities (see page 43).

What personal skills and attitudes are needed?

If you plan to become a fashion designer you should recognise most of the following traits in yourself:

- You are dedicated, motivated and ambitious. Fashion is such a competitive environment that you will need to be tenacious and thick-skinned to make it through.
- People comment on your clothing and personal style. You know what is in fashion, but do not follow it slavishly. Instead you prefer to use your knack for putting together outfits from a number of sources (eg charity shops).
- You have a highly developed sense of colour, being confident enough to go against the accepted wisdom if your combinations look good.
- You have commitment and self-discipline. Often the work is very self-directed, especially at college or if you set up your own business. You must be able to resist getting side-tracked or distracted and get on with the task in hand.
- You have flair and imagination, a tendency towards the dramatic and fanciful, the nerve to be different.
- You are talented at sewing and full of ideas. You often find that you imagine an item of clothing for yourself that you can't find in the shops. You also enjoy looking at fabrics and are comfortable with using paper patterns to make clothes.
- You have a bold and confident personality, you are an individual with strong tastes and preferences.
- You enjoy reading fashion magazines and find that you have a 'feeling' for what styles will be in fashion next year.

How competitive an area is it?

Competition for fashion design jobs is extreme. You have to be totally dedicated in order to succeed. It's worth thinking very hard about whether it is designing that you are attracted to, or some other area of fashion. For a purely design role, it is essential that you gear yourself towards a fashion design qualification by studying for the relevant exams. You may, however, be more suited to work on the manufacture or management side of the fashion/clothing industry. In this case, it will be beneficial to gain a qualification with a marked technical content such as clothing technology.

Haute couture and 'designer' store designers usually have degrees, but this does not preclude them from working in mass market fashion should they wish to do so. It is also worth considering if there is a type of clothing that you would be particularly interested in designing. If you hanker after hats, long for lingerie or have a feeling for footwear, you should be taking options in these at college, because a prospective employer will want to see evidence of the strength of your work in a specialist field.

Once you have left college you will have to rely on a number of factors to secure a job:

- A very strong and persuasive portfolio. The portfolio is the strongest evidence of your abilities. Be prepared to talk confidently about the pieces that you include and to put yourself across positively at interview.
- You should be seeking jobs not just by looking for advertisements in the trade press but also by making personal contacts. Work placements from college and

holiday jobs should have made you some useful acquaintances which you should cultivate. Don't be afraid to approach companies that you are interested in, even if they are not advertising jobs and you don't know anyone who can introduce you there. Write a letter to the design department or, even better, telephone and ask if you can come and talk to them. Once you are there you can take the opportunity to find out more about the company, to talk to the designers about how they got their first jobs and to show your enthusiasm. If you're lucky this may lead to an opening for you to ask for a job, or work experience.

- Don't confine yourself to looking for a full-time job – seek freelance work, commissions and work placements. Any work is good experience, paid or otherwise, and makes you a stronger candidate for the next job you pursue.
- If you cannot find work from other sources you may consider setting up on your own (see page 32).

If you really struggle to break into this competitive field as a designer, bear in mind that a design training has a wider scope of application in the fashion industry. You might find your training will lead to some of the jobs outlined on pages 4–13, and this will widen your career options.

What are the good and bad aspects of the work?

The glamorous side of fashion design is the one we are familiar with from the media portrayal of designers. The fashion world is very exciting, constantly changing and frantically paced. Designers work with interesting, creative people and the best-known ones mix with the rich and famous. All designers experience the satisfaction of seeing their ideas turned into a reality that people want to buy. The international travel and spectacle of the shows are part of what we perceive to be a jet-set lifestyle.

However, it is worth taking off the rose-tinted spectacles and being realistic for a moment. While there are very many rewarding and attractive aspects to working as a fashion designer, there is also a difficult side for which you should be prepared. As well as being an exciting world, it is a tough and competitive environment with high pressure, demanding deadlines to meet and clients who can be difficult. This can lead to a high stress factor and a very limited amount of time off. In the high street sector especially, you are likely to find that your designs are very much constrained by requirements in terms of budget and mass market appeal. Even if you reach a pinnacle of success in fashion design, it is by no means guaranteed to last, as there are always young ambitious people with fresh ideas pushing their way up the career ladder.

Where will I work?

Many fashion design jobs in Britain are based in London, especially central London. The UK's capital city is one of the world's most important fashion centres, being the home of highly prestigious design houses and spectacular shows. You may be familiar with the phrase 'London, Paris, New York' – these are the cities in which famous design companies will have a base, and the names will often appear on a designer label. Other world fashion centres are Milan and Hong Kong.

The importance of being 'where it's at' makes it tough for small companies. Commercial rents are very high in the capitals where most companies are based. Appearances are very important in the fashion world and it can be difficult if you are tucked away in a back street. Many companies have production centres outside London in order to minimise costs, locating factories in the Midlands and the North (areas that have been famous for textiles since the Middle Ages). Some British and American companies have also set up production centres in Third World countries because of the lower employment costs.

The size of individual companies varies enormously. Large mass market companies like Marks & Spencer, for example, employ thousands of staff in their clothing retail departments and factories. At the other end of the scale, the smallest haute couture companies can be made of up of just one or two very hard-working individuals. Because of the size of the mass market companies, there are many more job opportunities for designers, especially those with experience in specialist fields, such as sportswear, children's clothes or swimwear.

Who will I work with?

As a designer you will have varying degrees of contact with people doing the jobs outlined on pages 4–13. The closest contact you will have will be with the other designers in your company, with budget controllers and with the sample room staff. Although designing is a role that requires cooperation with other people, and good teamwork skills are essential for people who work in close creative groups, the designing itself is chiefly done alone. So be ready to work with a wide variety of others, but also be prepared to spend plenty of hours alone with just your sketch pad.

What will I earn?

The amount you earn very much depends on several factors, such as whether you are self-employed, what area of fashion you are working in, how successful your company is and how senior your position is. As a general guide, the most stable positions are in high street fashion companies where you are likely to receive a steady income and company benefits, such as a pension scheme. Have a look at the job vacancies in this sector, in your area and elsewhere (try your local paper, employment agency, the trade press and some national papers). See what salary ranges are offered for different levels of experience in different types of work.

What are the hours and holidays like?

Fashion design is not a nine to five job; it involves long and irregular hours, which can affect your home and social life. Particularly pressured times are usually in the run up to a high sales season, such as Christmas, a new season's deadline or a fashion show. Holidays will be fitted in as and when possible, rather than when it suits you. But as you will be working in a job which you are interested in and devoted to, you will probably enjoy the demanding nature of the work.

Will I meet the public?

Couturiers designing clothes for individual clients will have a high degree of contact with each one and sometimes personal friendships can grow out of conversations in the fitting room. Designers working with individual customers will have to learn how to deal with their aspirations tactfully and sensitively while at the same time steering their clients towards designs that will suit their shape and lifestyle.

Other designers will have to be in close contact with the profile or tastes of their customers, but are unlikely to meet them personally. If you thrive on contact with people, you could consider a career in fashion promotion, or in retail.

Are the prospects good for my career?

Competition for jobs is the toughest part of making a success of your career. You may increase your chances by moving into a specialist area, such as bridal wear, accessories or shoe design. Fashion design courses equip you for a wider range of jobs than just pure design work. If you find it impossible to get into designing as a career there are plenty of other options to consider which you may find just as rewarding. You really must be 100% committed, and if you are it is worth taking a shot at your dream. You may be the next Vivienne Westwood or John Galliano!

What about training at work?

In the past, your formal training would usually have been considered completed by the time you left college. Things have changed, for two reasons. First, the increasing application of new technology is creating far-reaching changes to society and patterns of employment. As a worker in the fashion and clothing industry, at whatever level, your usefulness will increasingly be tied to your willingness to upgrade your knowledge and skills to keep pace with the changes around you.

Second, job-related qualifications in the form of National or Scottish Vocational Qualifications (N/SVQs) are being developed all the time. These provide a framework of specialist workplace skills training at a series of different levels. Employees are able to develop their skills as they work, perform to a recognised standard, and achieve appropriate qualification. By this means, employers wishing to develop the supervisory and management skills of newly appointed fashion design graduates can encourage them to study in the workplace for an N/SVQ qualification. This makes sense, because there are some things that are very difficult to teach properly at college (like how to deal with your colleagues and how to act in a professional manner).

Not everyone, or course, wishes to be a student. As an alternative to college, you could take a Modern Apprenticeship route to gaining craft and technical skills in the clothing industry (see page 15).

Will I need other languages for the work?

As a working designer you are highly likely to travel abroad, and many people take up the opportunity to work in another country. If you are travelling to shows and fairs you are only likely to need basic phrases at the very most. If you work abroad, however, to make the best of your time and to live in another country comfortably you will need to be fairly fluent in the language. As the opportunities are most likely to arise in Europe, it may be well worth your while to study French, Italian or Spanish, and take advantage of language courses offered at college.

Will I be able to work overseas or travel for my job?

As previously mentioned, designers need to travel to fashion and fabric shows around the world and some choose to work abroad. Many designers find other European design houses more adventurous and innovative than British ones and so prefer to take up jobs there for a year or more. It is a very good career move as well as an enriching life experience to spend a year or two abroad. Aspiring haute couture designers often pursue this to their benefit, finding that it helps them to get better jobs in Britain when they return.

What are the recent developments in this area?

If you're interested in fashion you probably know that fashion and clothing design are constantly developing. Every season brings a new look, a new hemline, a new range of colours, and influences from different cultural and professional areas. Some fashions are actually revivals from previous eras, some borrow from other sources, such as the growing area of sportswear fashion – football replica jerseys and sport-derived footwear. New items have become fashion statements too – personal CD players that fit in your pocket, and mobile phones to match the colour of your accessories. These illustrate the blurring of boundaries between fashion and product design, and the way technology has influenced so many aspects of our lives – even the way we dress.

What impact has new technology had on this work?

As we have just seen, what we wear, and our essential accessories, have been affected by the huge technological advances we see all around us. But behind the scenes, in production and manufacture, changes have been even more dramatic. By the end of the 20th century, computer technology had already caused the biggest changes in clothing production in hundreds of years. The greatest improvements have been in computer-aided manufacturing (CAM) where technology is employed chiefly in the areas of pattern development, planning pattern layouts, cutting fabric and distribution. The technology that most affects fashion design is computer-aided design (CAD), and training in CAD is now provided on most college courses. By creating 3D images on screen, CAD can be used to aid planning and costing as well as pattern manipulation.

With production schedules and design details becoming increasingly computerised, fashion companies are able to manufacture clothes in direct response to sales reports, whereby if an item is selling well, more can be produced at short notice. This system is known as Quick Response and it is vital for enhancing a company's competitiveness.

The rise of specialist mail-order clothing companies has allowed detailed computerised databases to be constructed of customer likes and dislikes. This provides a wealth of valuable information to help the companies position themselves in a very competitive market. E-commerce, as a clothes-buying concept, is still in its infancy. However, analysts predict that many of us will soon be purchasing clothes through our personal computers and mobile phones. This will, in time, have tremendous implications for fashion retail.

Could I become famous?

Talent, drive and enthusiasm are not enough to make you a famous designer. The Katherine Hamnetts, Christian Diors, Giorgio Armanis and Ralph Laurens of this world have all had their share of lucky breaks. They have caught the mood of the moment and the eye of the media. Your chances of making it as big as one of these names are extremely slim. However, they do exist and the long odds shouldn't stop you going for it.

Could I work independently?

An option open to fashion designers is to set up their own label. There are a number of sources of funding available to young people setting up their own businesses, such as the Enterprise Allowance Scheme and the Prince's Trust. The risks are high and you are likely to struggle for a long time before you succeed or fail, so it takes lot of nerve and sacrifice if you are going to follow this course of action.

You could start up in a small way by hiring yourself a market stall and selling your own creations at a price that people can afford, or you could go a step further and set up a small, more up-market couture company creating one-off garments for individual customers on a commission basis.

Any new venture takes a great deal of planning and forethought, for which it helps to have a partner. It's also good to have partner to share the hard times and the successes. For advice about working for yourself, have a look at some books on self-employment and setting up a business.

How can I find out more about the work?

Finding out about the industry is something that you have already started to do by reading this book. Don't stop here – several other useful publications are available (see page 40 for some suggestions). Keep your eyes open for programmes on television about fashion, like the Clothes Show and one-off documentaries and features on art programmes. Speak to everyone you can about working in the fashion industry. Visit design studios and art colleges, chase personal contacts and speak to your school or college careers adviser.

Experience of working in the industry, paid or unpaid, will give you the best taster of what it is like, and you should try to find somewhere willing to take you on for a time. If you can't do so in a design studio, try the sample rooms, manufacturing plants and machining rooms.

What should I do now to prepare?

To be reading this book you must already be broadly interested in the fashion world. Look again at page 16 – do you fit the profile? If so you are probably already doing quite a lot to prepare. Here is a checklist of things you could be doing.

✔ **Be fashion conscious**

Be aware of what is happening in the fashion world. Magazines are a good source of information, especially *Vogue* and the trade press. You may be interested to read profiles of famous designers and general fashion books in the library (look in the arts section).

✔ **Start sewing**

You are probably already making your own clothes. This will usually involve using paper patterns to work from, but you shouldn't be afraid to experiment with your own ideas, initially using a cheap cloth (like cotton) to save on expensive mistakes. If you like the finished product you can always dye or paint it to make it more interesting, or make it up again in a better fabric.

You should be using a sewing machine. If you cannot borrow one from home or friends, there is probably one at school or college that you could use in lunch break or at the end of the day. It is important that you become comfortable and confident with a machine because at college you will be expected to handle industrial machines, which are faster and more powerful.

Spend time browsing in fabric shops and markets, examining the different qualities of the cloth, thinking about what different prints would be suitable for and sketch designs that come into your mind. (A word of warning: if you find yourself starting lots of projects and never quite getting around to finishing them, you may not be committed enough to stick out a further education course in fashion design.)

- ✔ **Talk to designers**
  Get into contact with working designers and ask to visit their studios. This will give you a better idea of what is involved in the work and what the working environment is like. Designers there may be prepared to chat to you about your ambitions, and may even be able to offer you some work experience helping out around the studio, fetching coffees and sandwiches, tidying up and so on. This doesn't sound very glamorous but it should be interesting and they may remember you when you start looking for holiday jobs while at college. You can also look for work experience in factories and sample rooms.

- ✔ **Visit colleges**
  Similarly, you should ask to visit your local colleges' art departments to get a feel of the atmosphere there and find out the emphasis and options of the available courses. It is essential to visit colleges' open days before applying to study at them. It is much easier to be interviewed by a tutor with whom you have already had an informal chat and they are more likely to offer you a place if they have seen evidence of prior interest. It will be expected of you to have a very clear idea of what the course involves.

- ✔ **Consider all your options**
  Find out if fashion design is really the career that you are most interested in pursuing. Think about the bad sides of the work and the pressures and think seriously about whether you can handle them. Consider other jobs that

interest you – would you prefer to go into those? You must be absolutely dedicated if you are to make it in fashion design, but dedication is not enough. You must also be talented. Try to assess your potential realistically. Seek opinions from people who will be honest with you about this – someone like your art teacher or any designers you meet.

✔ **Enter competitions**

One way in which young designers can attract attention to themselves is through fashion design competitions that are sponsored by fashion companies and the media. All fashion students are encouraged to enter these competitions as they can lead to commissions or full-time jobs. Don't be put off entering competitions if you aren't yet at college; it's good experience, and the originality of your design may win over the judges.

What courses and qualifications are available?

The following courses are the main routes in higher education, particularly for fashion design. See also pages 14–15 for information about courses up to 17/18, and vocational training.

Foundation

A one-year foundation course is usually entered after A-level studies. Funding is unlikely to be available to study these courses, so people tend to take them at a college near their home. Course tutors will usually expect most students to gain two or three A-level passes, but offers are based on your portfolio rather than on the attainment of certain grades.

Portfolio really means 'paper carrier' – ie a large folder – but the word now refers to the artwork held inside. This includes sketches, observational and working drawings, and completed works, or photographs of them. The purpose is to show the general scope and standard of your work, as well as the levels of your imagination and originality.

A strong portfolio would show your best work and give an idea of how you have progressed through your A-level course. Be prepared to talk about the examples that you include – the inspiration behind each piece, what you wanted to achieve, what problems occurred in the making and how you dealt with them, what is good and what is bad about the finished work. You are likely to be asked similar questions at interview, so think in advance about how you will handle them.

You must also show enthusiasm for and commitment to the course. Applications are made direct to the college, rather than through a central body. Once on to a course, you will find that it is aimed at widening your skills and experience in art. You will work in sculpture, painting, printmaking, textiles and 3D design. It is possible to specialise in your chosen area of fashion later in the year, but you may find that you prefer to pursue another area, for example photography or textiles.

Degree courses

The foundation course is an important entry requirement for most degree courses. Admission is through UCAS (The Universities and Colleges Admissions Service). If you are doing A-levels you will receive a copy of the UCAS booklet 'How to Apply' to help you through the application process (write to UCAS for a copy if not, address on page 44). Other course directories and useful publications will be available in your school or college careers library. Some Scottish institutions (eg Glasgow School of Art) should be applied to direct. Contact the individual institutions for advice.

Degrees generally fall into two categories: clothing technology and fashion design.

Clothing technology degree

This course is more oriented towards clothing production than design, covering all aspects of the manufacturing process. The design component of these courses is very small and it leads to careers in production management rather than fashion design.

Fashion design degree

Fashion design courses incorporate elements such as fashion drawing classes, design processes and projects that are based on commercial practice. Options are available for specialisation in various fields such as millinery, theatrical costume, children's wear, knitwear, history of costume and contouring. It is a good idea to bear in mind the job market when considering your options at college: you could think of taking technology, production management or computer classes. The last year is spent working chiefly in your specialist field and preparing for your degree show. The courses are very demanding, requiring real commitment and energy. You are assessed on all parts of your work, including your work processes. After your degree course you might want to pursue a postgraduate course or an extended study programme.

Higher National Diploma (HND)

Edexcel and Scottish Qualifications Authority (SQA) HND courses can be entered following a foundation course, with an art A-level or after completing a National Diploma or Advanced GNVQ (or Scottish equivalents). The core elements of the course cover the following areas: visual and design theory and techniques; the history and traditions of art and design; the business and professional aspects of design practice; verbal and written communication skills. As well as incorporating practical work skills in fashion design, the course also aims to improve general skills such as teamwork, problem-solving and planning.

It is essential to think very carefully about choosing courses and colleges to apply to. Visit each college you are interested in and ask about course options, facilities and work placements. Students on the course will have enlightening things to say as well as the course tutors. Be realistic about where you apply – there's no point applying only to colleges where you are not likely to win a place. Your tutors should be able to advise you about this.

What publications should I look at?

Periodicals and trade publications

The following is just a selection of the wide range of magazines and journals available in fashion, clothing design and textiles. These, and others, may be available for reference in careers libraries and public libraries, or on sale at newsagents. Also in newsagents you will find many of the fashion magazines, such as *Vogue, Elle, Harpers & Queen,* and so on. Check also the websites of professional bodies, as some publish their own journals and give suggestions for further reading.

- *Art & Design*, Wiley
- *British Style*, Beacon Enterprises
- *Clothes Show*, Focus Magazines
- *Costume* (annual), The Costume Society of Great Britain
- *Crafts*, Crafts Council
- *Creative Review*, Centaur Communications
- *Design* and *Designer*, Design Council
- *Design Week*, Centaur Communications
- *Drapers Record*, Emap Fashion
- *Fashion Weekly*, Mclaren Ltd
- *Menswear*, Emap Fashion
- *Textile Institute Journal*, Textile Institute
- *Textile Review*, Alain Charles Publishing

Career and professional

- AGCAS (Association of Graduate Careers Advisory Services) information booklets on art and design include: *Fashion and Textile Design; Your Degree in Art and Design*
- *Careers in Fashion* and *Careers in Art and Design*, Kogan Page

- *Creative Futures: A Guide to Courses and Careers in Arts, Craft and Design,* National Society for Education in Art and Design (NSEAD)
- *Designer Fact File*, British Fashion Council
- *Getting Jobs in Fashion Design*, Cassell
- *Your Creative Future,* The Design Council
- *Working in Fashion*, COIC

Further and higher education

- *Art and Design Courses*, in the Complete Guide series, UCAS/Trotman
- *CRAC Directory of Further and Higher Education*, Hobsons
- *Degree Course Offers*, Trotman
- *Entrance Guide to Higher Education in Scotland*, COSHEP
- *Getting into Art and Design*, Trotman
- *Guide to Higher Education for People with Disabilities*, available from Skill

Student finance

- *Financial Support for Students*, a Department for Education and Employment guide, annual (also available on the DfEE website: www.dfee.gov.uk); *Student Support in Scotland* from the Student Awards Agency for Scotland (SAAS); and *Financial Support for Students in Higher Education* from the Department of Education for Northern Ireland (DENI).
- *Sponsorship and Funding Directory,* Springboard Student Services, CRAC (published annually).

Which addresses will help me?

Arts Councils:

Arts Council for Wales
Holst House, 9 Museum Place, Cardiff CF1 3NX
tel: 01222 349711

Arts Council of England
14 Great Peter Street, London SW1P 3NQ
tel: 020 7333 0100
website: www.artscouncil.org.uk

Arts Council of Northern Ireland
Stranmills Road, Belfast BT9 5DU
tel: 028 9066 3591

Scottish Arts Council
12 Manor Place, Edinburgh EH3 7DD
tel: 0131 226 6051
website: www.sac.org.uk

Association of Illustrators
81 Leonard Street, London EC2A 4QS
tel: 020 7739 8901
website: www.aoi.co.uk

Association of Photographers
81 Leonard Street, London EC2A 4QS
tel: 020 7739 6669
website: www.aophoto.co.uk

British Clothing Industry Association
5 Portland Place, London W1N 3AA
tel: 020 7636 7788
email: bcia@dial.pippex.com

Also at this address (5 Portland Place, W1N 3AA):
British Knitting and Clothing Export Council
British Apparel Centre
British Footwear Association (tel: 020 7580 8687)

CAPITB Trust
80 Richardshaw Lane, Leeds LS28 6BN
tel: 0113 227 3345
website: www.careers-in-clothing.co.uk

Chartered Society of Designers
32–38 Saffron Hill, London EC1N 8SG
tel: 020 7831 9777
website: www.csd.org.uk

City & Guilds of London Institute
1 Giltspur Street, London EC1A 9DD
tel: 020 7294 2468
website: www.city-and-guilds.co.uk

COIC (Careers and Occupational Information Centre)
PO Box 298a, Thames Ditton, Surrey KT7 0ZS
tel: 020 8957 5030
website: www.dfee.gov.uk

Crafts Council
44A Pentonville Road, London N1 9BY
tel: 020 7278 7700
website: www.craftscouncil.org.uk

Department for Education and Employment
Student Support, Mowden Hall, Staindrop Road,
Darlington, Co Durham DL3 9BG
tel: 01325 392852
website: www.dfee.gov.uk

Design Council
34 Bow Street, London WC2E 7DL
tel: 020 7420 5200
website: www.design-council.org.uk

Design Museum
Shad Thames, London SE1 2YD
tel: 020 7403 6933
website: www.designmuseum.org

Edexcel Foundation (BTEC)
Stuart House, Russell Square, London WC1B 5DN
tel: 020 7393 4444
website: www.edexcel.org.uk

London College of Fashion (The London Institute)
20 St John Princes Street, London W1M 0BJ
tel: 020 7514 7407
email: enquiries@lcf.linst.ac.uk

National Society for Education in Art and Design (NSEAD)
The Gatehouse, Corsham Court, Corsham SN13 0BZ
tel: 01249 714825
website: www.nsead.org

Scottish Qualifications Authority (SQA)
Hanover House, 24 Douglas Street, Glasgow G2 7NQ
tel: 0141 248 7900
website: www.sqa.org.uk

Skill (National Bureau for Students with Disabilities)
Chapter House, 18–20 Crucifix Lane, London SE1 3JW
tel: 0800 328 5050
website: www.skill.org

Textile Institute
St James' Building, Oxford Street, Manchester M1 6FQ
tel: 0161 237 1188
website: www.texi.org

UCAS
Rose Hill, New Barn Lane, Cheltenham GL52 3LZ
tel: 01242 227788
website: www.ucas.com